Kattabomman's Trick

written by
Sowmya Rajendran

illustrated by
Niveditha Subramaniam

All rights reserved. No part of this publication may be reproduced or distributed in any form or by any means, or stored in a database or retrieval system, without the prior written permission of the publisher.

Kattabomman was a turtle. He did not like it when people called him a 'tortoise' because you see, he wasn't one.

He was also the laziest turtle ever.
If he could help it, he would not move from his favourite place behind the white shell with brown speckles at all.

Whenever Kattabomman smiled, his dimple would show like a little sun peeking out of the clouds.

He really was a very nice looking turtle!

One day, Kattabomman decided to play a trick on his friends in the fish tank, Kumudham [the crab], Gunavathi [the goldfish], Uncas [the swordfish], Pran [the prawn], and Forgot [the fish].

"I'm going to disappear for a while!" he said to the mermaid doll and winked.

Kattabomman pulled his head inside his shell. This is a magic trick that only turtles can do and Kattabomman was very proud of it.

Actually, tortoises can do it too, but Kattabomman did not like it if people told him that.

For a long time, the fish tank was very quiet. Nobody noticed that Kattabomman had disappeared.

But suddenly, Pran, the prawn, who was always the first to notice things, asked, "Where is Kattabomman?"

Gunavathi looked around in surprise. "Where is he?" she wondered.

Uncas, the swordfish who liked adventures, immediately said, "It's a mystery! It's a mystery! We have to find him! Come on, hurry up, let's go, let's go!"

Pran smiled at Uncas and said, "We should, perhaps, make a plan."

"A plan? Why do we need a plan?" asked Gunavathi who always liked *knowing* things.

"We must *first* draw a map of the fish tank to make sure that we search E-V-E-R-Y-W-H-E-R-E. Otherwise we might miss a place and not find Kattabomman. That's why!" explained wise Kumudham.

Gunavathi had to hold down Uncas who was excitedly doing ten flip-flops a minute.

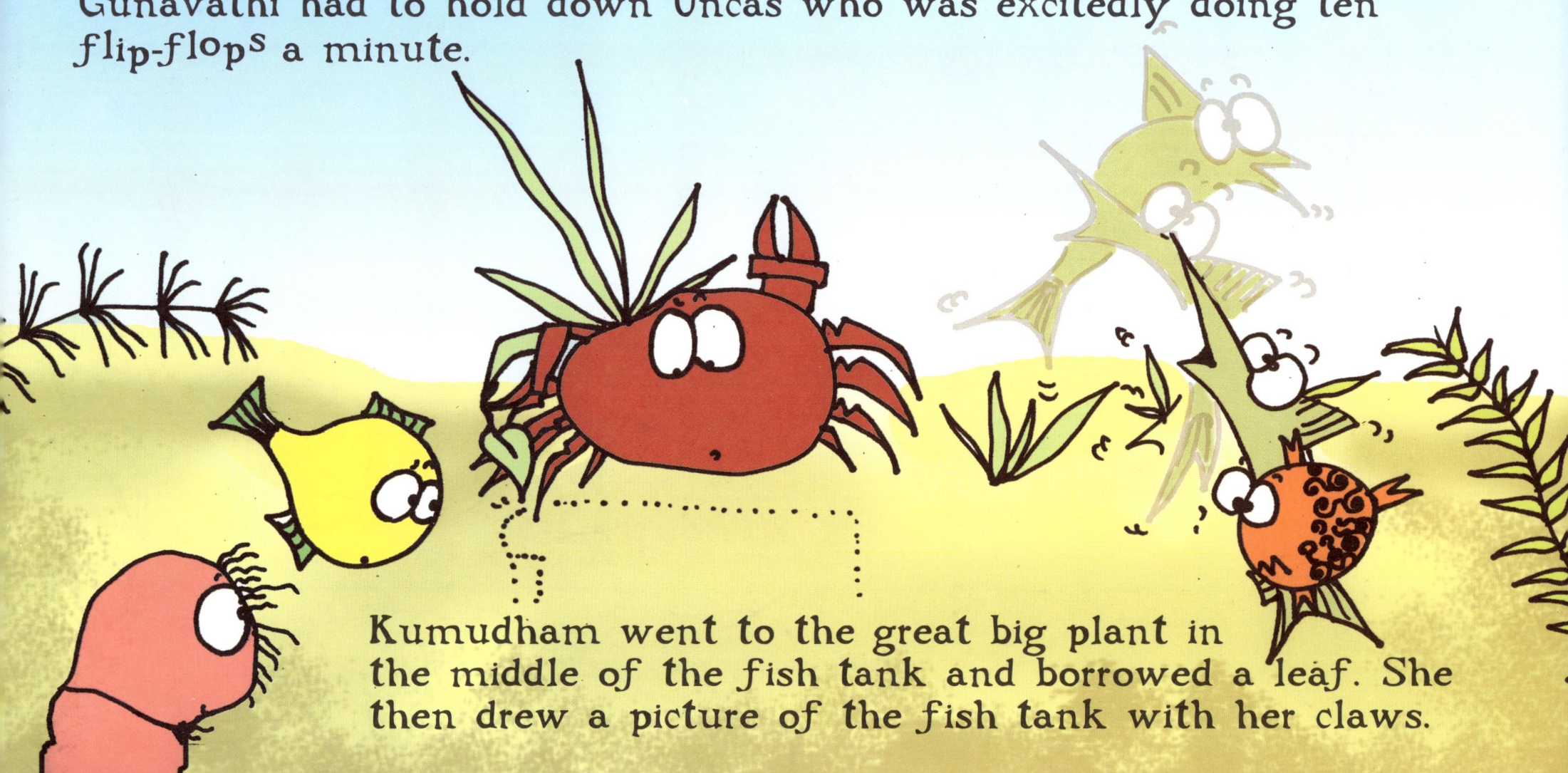

Kumudham went to the great big plant in the middle of the fish tank and borrowed a leaf. She then drew a picture of the fish tank with her claws.

Meanwhile, what was Kattabomman doing?

Kattabomman had been sitting inside his shell and listening to the excitement outside.

He was feeling so happy that he started daydreaming. He dreamt about food.

Lots and lots of it.

He dreamt so much that he fell asleep.

Outside his shell, the fish tank was full of activity.

Kattabomman's friends searched and searched. They searched E-V-E-R-Y-W-H-E-R-E, but they could not find Kattabomman.

Instead, Gunavathi found a new shell behind the white shell with brown speckles.

"How did this new shell get here?" Pran asked.

"It is a rather nice shell," said Forgot.

"It's a magic stone and I am sure there is a genie inside it!" shouted Uncas in excitement.

Kumudham, who was very wise, said, "We must be careful. Kattabomman is missing and there's a new shell here. We must find out what is happening."

Forgot shivered in fear and asked, "Do you think he is okay?"

But Uncas was not listening to any of this. He was busy bouncing on the new shell and waiting for the genie to appear any minute.

Inside his shell, Kattabomman began to stir. What was this thud-thud noise on his back?

Thud!

Thud!

Thud!

...jumped Uncas on his back.

Kattabomman woke up. He felt really hungry after his dream. He popped his head out and asked,

"Anybody got any food? I'm starving!"

Uncas did three flip-flops all at once in surprise.

"It's the genie!" he shouted with joy.

"No! It's Kattabomman!" said Pran happily.

"You had disappeared for so long that I nearly forgot all about you!" said Forgot smiling.

Kumudham and Gunavathi did a merry little dance together.

Suddenly, everybody felt as hungry as Kattabomman.

They decided to have a picnic behind the white shell with brown speckles because Kattabomman did not want to go anywhere else.

"Let's stay here, right here!" he said.

You see, he was the laziest turtle ever, and if he could help it, he would not move from his favourite place at all!

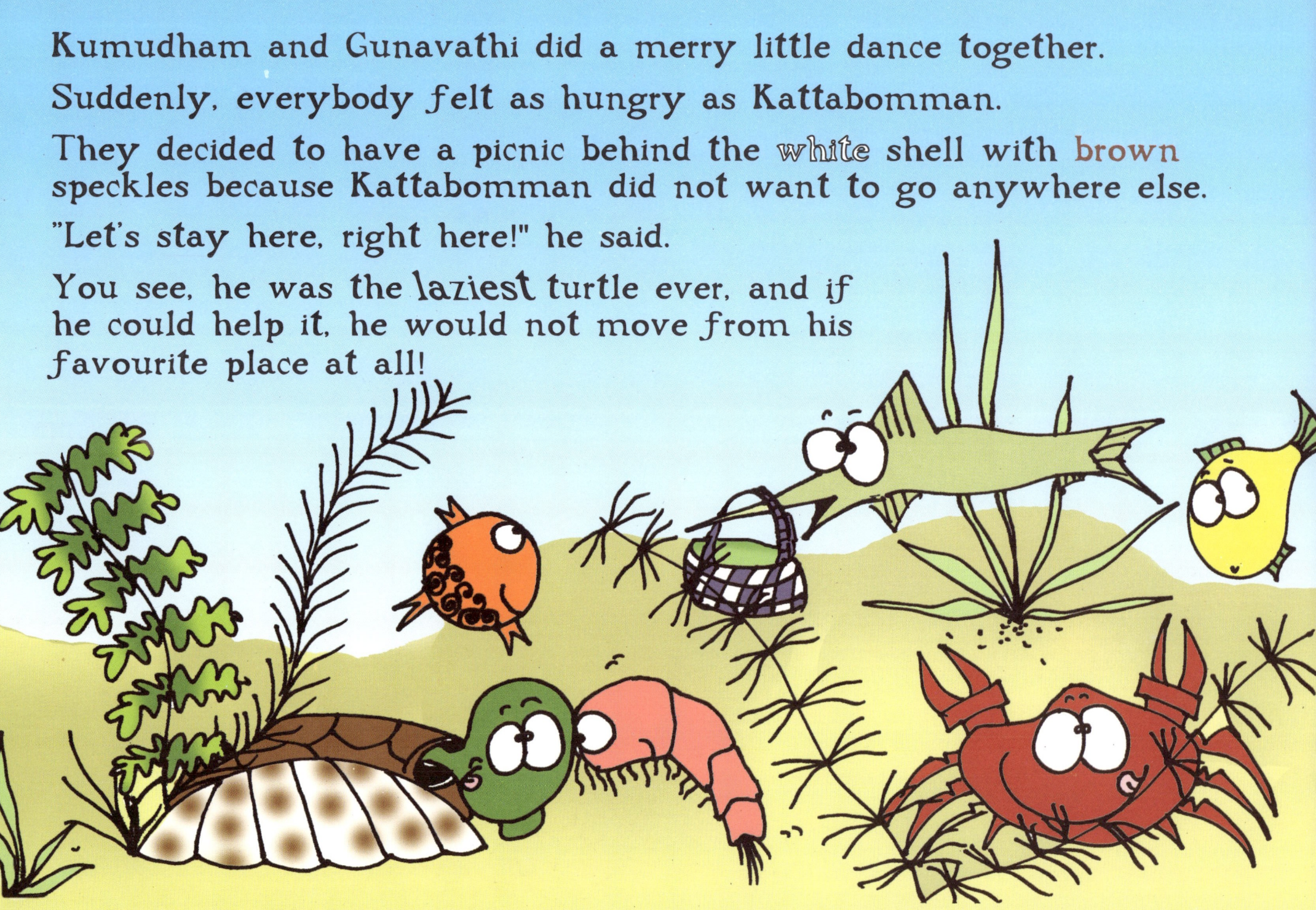